This booklet is based on a trip we made from 11th June to 2nd July 1991, (mainly) at Varanger Fjord, Norway). It describes where, and how, we found birds such as Terek Sandpiper, Yellow-breasted Bunting, Greenish Warbler, Arctic Warbler, Booted Warbler, Blyth's Reed Warbler, Little Bunting, Broad-billed Sandpiper, Jack Snipe, Pine Grosbeak, River Warbler, White-tailed Eagle, Tengmalm's Owl, Smew and, even Two-barred Crossbill. It also gives details of where to search for some of the birds we missed such as Gyr Falcon, Pygmy Owl, Spotted Eagle and Red-flanked Bluetail.

Throughout our trip we kept meeting birders from many countries who were always willing to be helpful. Generally, however, we had found more than they had, so I was frequently drawing sketch maps like the ones shown here. I hope they found them useful and I hope you will too.

Have a great time in Finland.

                        Dave Gosney,
                        September 1991

Acknowledgements

Thanks to Ian Lewis, William Smith and C.R. Brooks for their permission to use information from their own trip notes, provided by Steve Whitehouse's Information Service. Thanks also to Keijo Vesanen and all the other helpful birders we met. Lastly, thanks again to Ann Whorton for the typing, Kim Waterhouse for the printing, Broadleaf for the cover and my parents for their tolerance.

Published by GOSTOURS, 29 Marchwood Road, Sheffield, S6 5LB
Printed by Kim Waterhouse
c 1992 D. Gosney. No photocopying please
ISBN 0 9517920 4 0

Oulu

Attraction

Oulu is most famous for two rare breeding birds: Yellow-breasted Bunting and Terek Sandpiper. The marshes of Liminka bay are, however, worth visiting in their own right, at least in spring when they have an excellent collection of ducks, waders, gulls and cranes.

Getting there

Oulu is a major town near the head of the Gulf of Bothnia with flights available to and from Helsinki.

Notes

1. Oulu oilport is the traditional site for Terek Sandpiper. (See page 4.)

2. Kerseleenlahti is the most convenient site for Yellow-breasted Bunting, being so close to Oulu and en route to the airport. An observation tower here overlooks the estuary. (See page 5.)

3. At Liminganlahti the most easterly bird tower was described by Lewis as the best of the three for birds but it can only be reached by a 2 km walk along a boggy track.

4. Just beyond Liminganlahti Information Centre is an excellent bird tower and the most reliable spot for Yellow-breasted bunting. (See page 6.)

5. The most Westerly tower at Sannanlahti may have birds which cannot be seen from the others. (See page 6.)

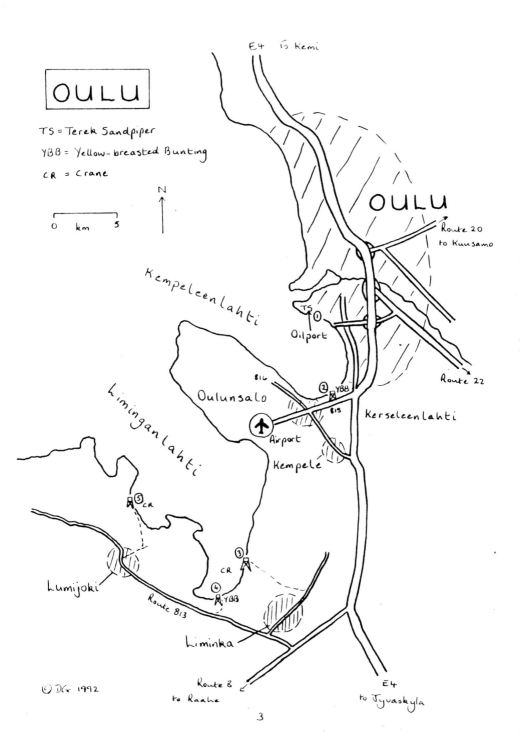

Oulu Oilport

Getting there

Drive south along the E4 through Oulu until you see signs for a right hand turn to the airport. Head towards the airport but at the next junction (traffic lights) turn right again back towards Oulu. After 2.2 miles (3.5 km) take the turning to the left just after a race track. The slip road leads to a T-junction; the road to the right follows the edge of the bay all the way to the oilport. (This is not the shortest route but it's the easiest to describe!)

Notes

The oilport lagoon where I have previously had Caspian Terns and Temminck's Stint has now been filled in. Two Finns told me that it wasn't worth looking here and took me to a 'secret site' further North where a pair of Terek Sandpipers summered in 1990. Sadly, this too was greatly changed and we saw little. However, a 'last hope' visit to the oilport proved that there were still a few pools with Snipe, Redshank, Wigeon etc. and a reedbed with a pair of Marsh Harriers. In a pool which was little more than a roadside puddle we did find a Terek Sandpiper (marked TS).

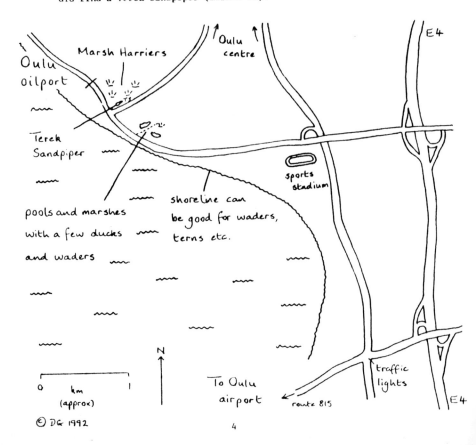

© DG 1992

Kerseleenlahti

Getting there

From the E4, south of Oulu, follow signs to the airport. Once off the E4 go straight on at the next traffic lights and then look for a small track to your right, 1.5 km (0.9 miles) from the lights, signposted Kerseleenlahti. Drive down this track to a car park then walk to a raised tower.

Notes

The tower overlooks an estuary area with plenty of ducks, Great Crested Grebe, Whooper Swans, Marsh Harriers etc. The walk to the tower is through willow scrub which looks ideal for Yellow-breasted Bunting although we only found Scarlet Rosefinch here. This is presumably the "bushy area to the right of the road to the airport" described by Gooders although it is more like 2.5 km not 0.5 km from the E4. The buntings presumably do occur here although Liminganlahti Information Centre is a much better bet.

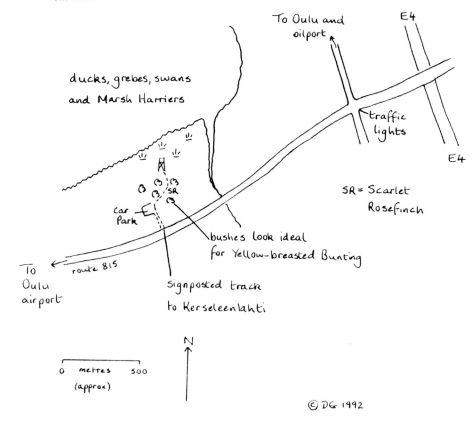

© DG 1992

Liminganlahti (Liminka Marshes)

Getting there

Follow the E4 south from Oulu until the right turn along the E8 towards Raahe and Ylivieska. Just a few km along the E8 is another right turn signposted to Liminka and Lumijoki (the E813). All the observation towers are to the North of this road.

1. We didn't visit this tower which is found by turning right (North) from the E813, into Liminka town. Go straight through the town and then look out for a signpost to Lintuntuuri (Bird Tower).

2. To find the Information Centre don't turn off the E813 into Liminka but continue towards Lumijoki. About 5 km after Liminka look for a sign saying "Liminganlahti Opastuskeskus" which of course means "Liminka Marshes Information Centre". This will direct you down a track to your right where, by keeping left, you will come to the centre and a car park. Walk down the left side of the centre to the bird tower.

   From the car park listen for Ortolan Bunting in the nearby fields. The willow bushes between the centre and the tower have several pairs of Yellow-breasted Bunting. A painting of one indicates a good place to see them from! The song is much more yodelling than other buntings. The phrase "Doodle-oodle eedle-oodle-oo" is not a perfectly accurate transcription of its song but it gives an idea of its lilting quality. Listen too for Scarlet Rosefinch singing "Pleased to meet you" even while still in juvenile plumage. The observation tower was superb on 12th June but somewhat disappointing two weeks later. In mid June there were 100+ Wigeon, Whooper Swan and Spotted Redshank, 30 Greylag, 60 Shoveler, 2 Pintail, 2 Garganey, 50 Merganser, 20 Goosander, 60 Curlew, 20 Ruff, 30 Little Gull and 6 Black-tailed Godwits in front of the hide. 6 Marsh Harriers, a male Hen Harrier and a Short-eared Owl flew around and in the distance were a flock of Cranes and a party of 6 Red-necked Phalaropes.

3. The most westerly tower offers similar birds but not in such profusion although we did have 60+ Cranes at close range plus 6 Little Gull, 10 Pintail and a Long-eared Owl flushed from bushes by the car park. To find this tower, turn right in Lumijoki, just past the church and a stream, along a road signposted to "Keri" and "Lintuvesi". Ignore the right fork into the hamlet of Keri and follow the track for a further 3.1 miles to a car park on the left. Walk a little further down the road to find a track to the tower.

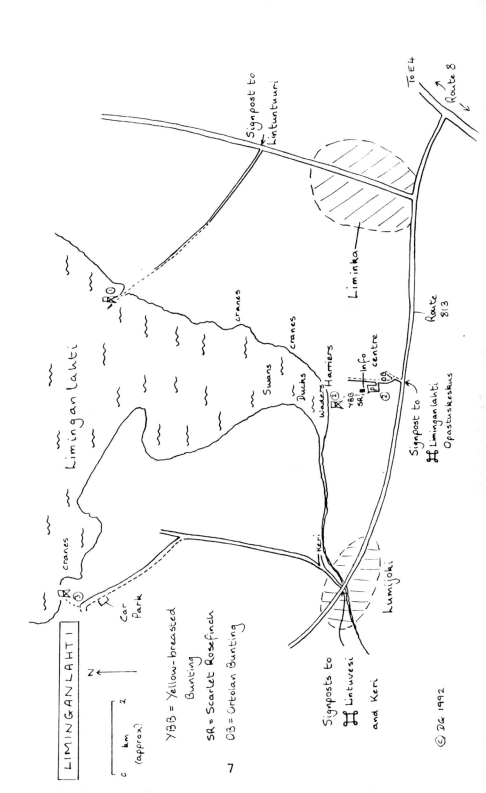

Kuusamo

Attraction

The best areas of boreal forest in Finland are found around here. In 1991 we found Little Bunting, Rustic Bunting, Arctic Warbler, Greenish Warbler, Pine Grosbeak, Three-toed Woodpecker, Black Woodpecker, Hazelhen, Capercaillie, Black Grouse, Smew, Lesser Whitefronted Goose, Waxwing, Siberian Tit, Siberian Jay, Parrot Crossbill and Two-barred Crossbill. Up to 30 singing Red-flanked Bluetails have been recorded in one season.

Getting there

Kuusamo is on the E20, 216 km north east of Oulu.

Notes

1. Within Kuusamo town is a large lake called Toranki. Here there are Little Bunting, Rustic Bunting, Smew, Goosander and Goldeneye. We also had Lesser Whitefronted Geese here in 1991. (See page 10.)

2. About 30 km north of Kuusamo is the famous Vaaltavaara ridge where Bluetails are usually found in summer. Whilst searching for these we found virtually every other sub-arctic forest bird you could expect. (See page 11.)

3. Further North still is the Oulanka National Park - an area of spectacular scenery where forest birds certainly occur but may be hard to find. (See page 14.)

4. The gorge at Ristikallio is quoted as a Red-flanked Bluetail site by Gooders. Our brief stop here was unproductive.

5. This shows the route of the Karhunkierros or Bear Ring Trail, a 50 mile footpath from Ristikallio to Ruka. Andy Wildin walked this in 1991 but found little apart from Siberian Jays and Dippers.

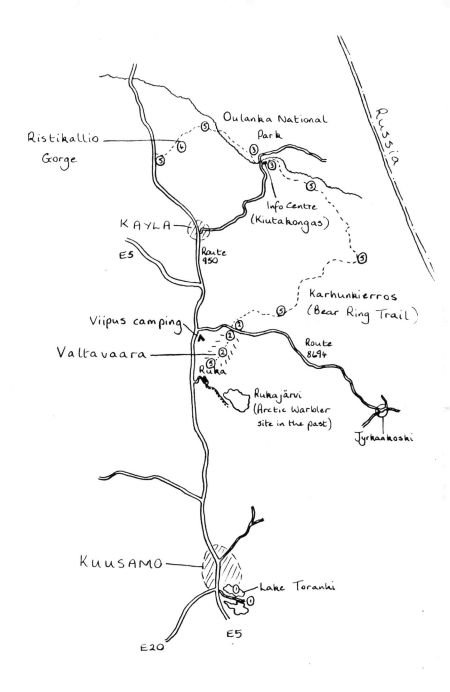

Kuusamo-Toranki Lake

Getting there

Approaching Kuusamo from Oulu, you arrive at a major road junction just south of the town. Here, on the right, is a large, well-signed, information centre. Drive towards this building but, instead of entering the car park, carry straight on. This road, which is called Torangivaital, follows a ridge of land between two lakes. Park at a point where both lakes can be seen from the road and walk from there.

Notes

1. Little Buntings and Rustic Buntings can be seen or heard in the birch scrub between the road and the two lakes. Our 1991 sightings are shown on the map. Little Bunting song is lighter and more elaborate than the local Reed Buntings whose song is also more extravagant than our own birds. Rustic Bunting song sounds like rather abrupt phrases of Dunnock song.

2. Looking down the road from the Little Bunting area you will see signs to a bird tower or Lintuntuuri, illustrated with a Whooper Swan. The path down to the lake on the left is not always clear but keep going in what appears to be the right direction and the lakeside tower will come into view. The first three steps to the tower were missing in 1991 so less gymnastic birders may have to miss out here.

   From the tower you get an excellent view of the lake. We made several visits and always found Smew (up to 6) plus lots of Goosander and Goldeneye. Lesser Whitefronts flew over on one occasion and another Rustic Bunting was seen by the path.

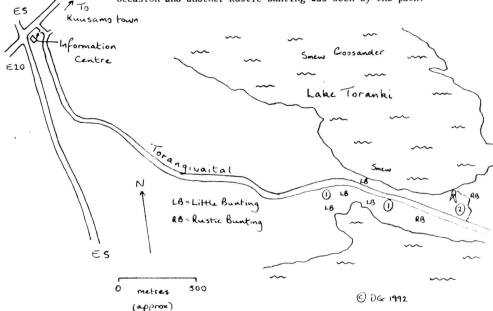

© DG 1992

Vaaltavaara

This is one of the most exciting birdwatching areas in Europe and yet for much of the time it seems devoid of birds. Patience, luck and a willingness to birdwatch all night are essential if you want to make the most of this place.

Getting there

Follow the E5 north of Kuusamo until, 5 km beyond Ruka, you see a road to the right signposted Jyrkankoski. Some very good birds can be seen around this junction but to find the best birds you must take this minor road for 4.4 km. After a lake on the right, the road rises over a ridge and a car park appears on the left. Park here.

Notes

1. From the car park, walk back down the road looking for a footpath to the south marked Karhunkierros (Bear Ring Trail). Even from the road you should be listening for Greenish Warbler, Arctic Warbler and Red-flanked Bluetail. Both warbler songs are unmissably loud; Greenish has a rich flowing warble; Arctic makes an almost mechanical buzzing noise like Cirl Bunting crossed with River Warbler but even louder. Tapes of Red-flanked Bluetails suggest that the song has a rhythm which goes something like "if-it-doesn't-fit, put-it-in-the-bin" with great emphasis on each note in the second half. However I haven't yet proved that this is what the real birds sound like.

2. The Bear Path Trail winds steeply up to the 'first hill' along the ridge. It was here that we located our family of Two-barred Crossbills in 1991. The call is obviously that of a crossbill sp., but is much lighter than the others with more than a hint of Redpoll about it. This is a good place to listen for Bluetails. Hazelhen are frequently heard where the path leaves the road.

3. On the 'second hill' along the ridge we twice found Pine Grosbeaks. Their call also sounds like a crossbill but is unmistakeably double-barrelled. The fluttering sound made by their wings may help you to locate a bird if it flies behind you. Don't just look in trees; they spend quite a bit of time on the ground amongst bilberry. Siberian Jay and Three-toed Woodpecker were also found here.

4. Further on, a sign to a shelter is worth a slight diversion. Siberian Jays love the area around the picnic tables and Black Grouse lek behind the marsh. We had another Arctic Warbler singing on the slope down to the big lake to the North, on 12.6.91 - the first at Kuusamo that year.

    We did walk further along the ridge but found nothing new apart from a Short-eared Owl.

5. From Valtavaara you can see a forest track between two lakes. A Finnish birder heard a Two-barred Crossbill here. We only found a pair of Hazelhens in the young plantation as shown.

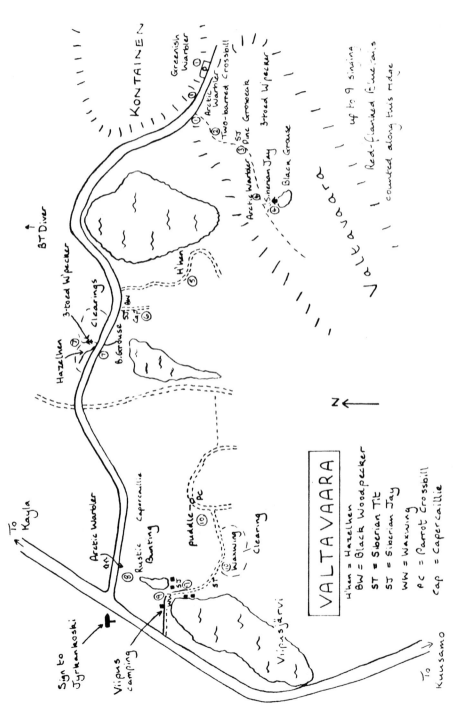

6. Martin Daniel had 3 male Capercaillies just 100 metres down this track. We couldn't find these but we did have great views of a Black Woodpecker.

7. The boggy birch scrub, North of the road where a stream runs through, is another site for Hazelhen. Behind here are two clearings half-separated by a line of pines. We twice saw Three-toed Woodpecker in this line of pines. Black Grouse were also seen from the roadside near here, sitting in trees.

8. Between the small lake and the roadside, another Arctic Warbler was heard and a pair of Rustic Buntings were breeding.

9. The campsite at Viipus camping was frequently visited by Siberian Jays and Waxwings. Greenshank and Green Sandpiper occurred on the lakes.

10. By walking through the campsite and following the track (look out for Siberian Tit) you will come to a clearing where we found two pairs of Waxwings and two pairs of Greenshank. The track leads back into woodland where, in 1991, a puddle had formed which attracted Parrot Crossbills.

Oulanka National Park

Getting there

From Kuusamo drive North for 36 km then turn right to Kayla on route 950. In Kayla, follow signs for Kiutakongas or Oulanka National Park. These will take you east on the 8693. About 15 km from Kayla look out for the Information Centre and car park on your right as you descend into a valley.

Notes

We followed the Luontopolku (round-walk) through beautiful-looking woodland but saw virtually nothing. We did however hear Black and Three-toed Woodpecker as shown on the map. Mid-afternoon was obviously not the best time to visit. Smith also had Dippers on the spectacular river and Siberian Jays at the picnic site; Lewis had Capercaillie and Hazelhen here.

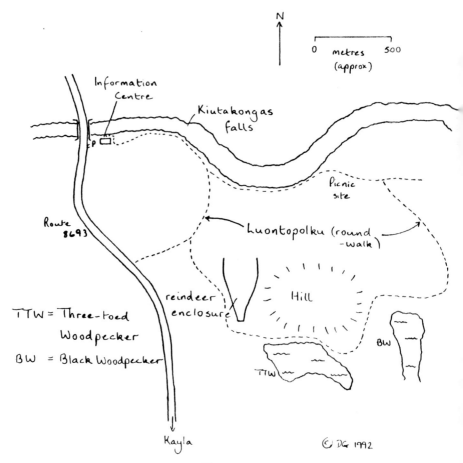

Pyhatunturi

Attraction

A good area of woodland (better than most around Inari) and marsh. Tengmalm's Owls seem very easy to find here.

Getting there

Pyhatunturi is a ski centre found by turning west from the Kemijarvi-Sodankyla road just south of Pelkosenniemi. It is well signposted. Park in the car park to the left of the road before reaching the Information Centre car park.

Notes

1. As you begin the nature trail, the path forks, following two similar routes which rejoin 100-200 metres later. Just before they rejoin, in the woodland to the right we found a Tengmalm's Owl nest in an old Black Woodpecker hole. We could tell it was occupied because lots of fluffy white feathers stuck to the trunk below the hole and, less conspicuously, because inside was the head of a young Tengmalm's.

2. Further on, where wooden steps head down into a valley, Lewis et al found an adult Tengmalm's Owl at a nest site.

3. The marsh was reasonably productive. With patience we found a Red-necked Phalarope, a Spotted Redshank and a Whimbrel as well as the more conspicuous gulls and Arctic Terns.

4. The rest of the walk yielded 18 species (pretty good for midday in these forests - around Inari we struggled to get 6!) including Waxwing, Siberian Tit, Crossbill and Raven. Just after the loop trail crosses another track is a point where Lewis et al found three juvenile Tengmalm's Owls.

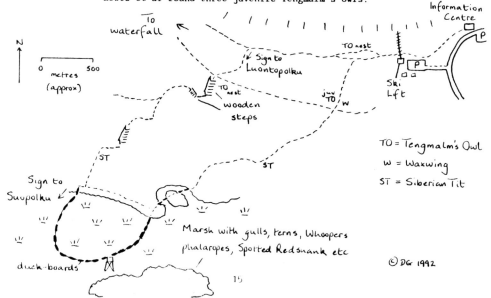

Karigasniemi

Attraction

A superb marsh nearby has lots of breeding waders including Broad-billed Sandpiper and Jack Snipe. Mount Ailigas has Ptarmigan, Dotterel and, reportedly, Arctic Redpoll. Rough-legged Buzzards are numerous and White-tailed Eagle and Gyr Falcon frequently seen. Red-flanked Bluetail has bred not far away.

Getting there

Karigasniemi and Utsjoki are border towns in the north west of Finland next to Norway. The road connecting them on the Finnish side of the border is perfectly good.

Notes

1. The famous area of marsh is between the 13 and 14 km posts from Karigasniemi. Some Scottish bird photographers we met there were finding amazing numbers of nests including several each of Jack Snipe and Broad-billed Sandpiper and maybe 20 Reeve nests. We had these species plus Spotted Redshank, Scaup, Red-necked Phalarope, Hen Harrier, Rough-legged Buzzard, Whimbrel etc. as shown on the map. We tramped across areas of marsh but in fact all the best birds were seen by the road. A short walk to the pool with Arctic Terns would suffice.

2. A pool to the south of the road about 10 km from Karigasniemi is worth a stop to enjoy lots of waders, especially early in the morning. Twenty-plus Red-necked Phalaropes were a highlight here.

3. A track 5 km South-east of Karigasniemi leads towards Mount Ailigas. Park at the barrier after 2 km and walk to the summit. In wind and driving rain we had only Snow Bunting and Lapland Bunting but others have had Dotterel and even Arctic Redpoll in "little valley near top" (Brooks).

4. Karigasniemi village has a few facilities including supermarkets, youth hostel, campsite and restaurant. Listen for Waxwings. To find the village turn right at the bottom of the hill (signposted Utsjoki) otherwise you'll run into the customs officers at the Norwegian border.

5. The whole valley from Karigasniemi to Utsjoki is excellent for Rough-legged Buzzards (we counted 16). A brief stop at the car park called Pahtavaara produced good views of White-tailed Eagle. This is a good area for Gyr Falcon but we only saw a Merlin. It would be surprising if the woods along this valley didn't have Arctic Warbler though we heard none.

6. There is a second Mount Ailigas at Utsjoki. It is also located by driving towards a radio mast as far as a barrier then walking onto tundra. Lewis et al had Long-tailed Skua and a distant Gyr Falcon here as well as Snow Bunting and Ptarmigan.

7. Some Italian birders we met told us that Red-flanked Bluetail had been seen, in previous years, in woods near Kevo National Park.

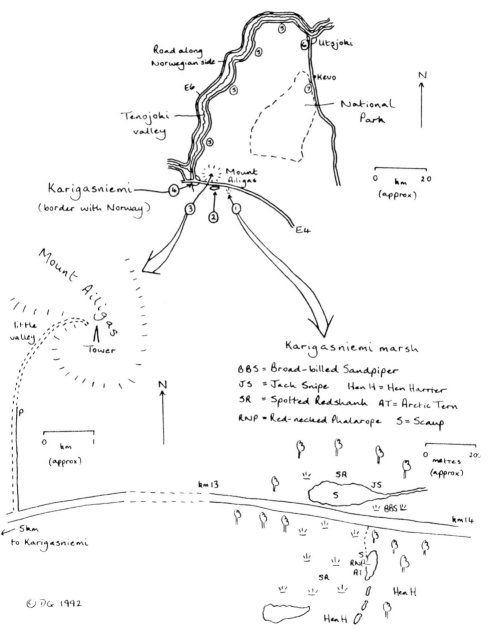

Inari-Ivalo-Vuotso

Attraction

Traditionally regarded as the prime area for Finnish birds of forest, lake and marsh with a few good hill species too. However, we found the forests to be very disappointing compared with those around Vaaltavaara and species such as Little Bunting, Smew and Jack Snipe are much easier to find elsewhere. The hill of Kiilopaa is worth a visit though.

Getting there (see map overleaf)

Notes

These notes are a compilation of everything I've managed to find on the area. As you will see, we didn't have much success here but then, having already seen these species, we didn't look all that hard.

1. Lewis et al had Waxwing, Pine Grosbeak and Three-toed Woodpecker around a wooded ridge North of Inari. Brooks had Siberian Tit and Waxwing in this area but Smith was unable to find a site which fitted Lewis' description and saw little. We found a lay-by 12 km (8 miles) North of Inari which seemed like the right place but the woods here looked unexciting and produced only Waxwing and Greenshank.

2. Smith suggested that a reserve to the south of Inari (Juntaan Luonnonkoitomesta) looked better for woodland birds. We found the car park on the left 4.3 km from Inari on the road to Kittila but, apart from the obliging Siberian Tits at a nestbox only 50 metres from the car park, we thought this area was poor. Our best birds were Waxwing and a Rough-legged Buzzard.

3. The area around Ivalo is said to be good for Little Bunting and Jack Snipe but Smith, unable to locate the sites for these described by Gooders and Lewis, found nothing of interest here despite much searching. Brooks however, with the help of a local contact, Hra. H. Karhu, sussed out three sites for Little Bunting

    a. Next to the Murmansk junction, a marsh at the end of a lake.

    b. En route to Veskonicmi where a rivermouth becomes visible. (This may be where Lewis had Jack Snipe - if so, it's on the left after 4 km.)

    c. En route to Koppelo, long marsh on left after about 5 km.

    Even after 2 days Brooks found Little Bunting at only one site, the last one. We didn't bother to look.

4. Some woods here, 2.5 - 4 km from the Ivalo-Nellim road were reckoned by Smith to be the best they found. They had Waxwing Siberian Tit and Siberian Jay here.

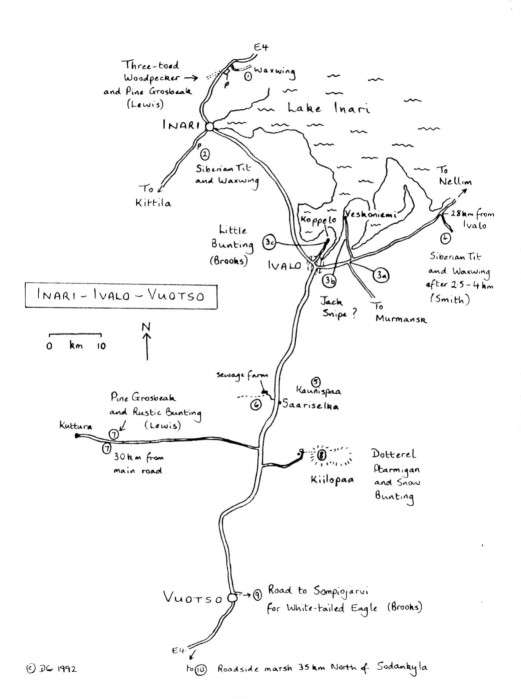

5. Kaunispaa is reckoned by Gooders to be a good hill for birds but Smith found it to be overpopulated with tourists. Kiilopaa (note 8) is a much better bet.

6. Saariselka Sewage Ponds may be worth a visit. We had Arctic Tern, Greenshank, Ringed Plover and several ducks here. To find them, turn west in Saariselka opposite a sign to the Information Centre, and drive up a track past a blue and yellow building until the pools come into view on the right. This track leads to woodland which is said to be a site for Capercaillie (especially beneath and beyond some power lines). Like Smith we found nothing of interest in these woods.

7. The best looking woodland we found in this area was about 30 km along the road to Kuttura. Here the woods were damp with mainly spruce and birch. However, in miserable weather, Siberian Tit and Wood Sandpiper were the best birds we found. Smith found only Three-toed Woodpecker here and Brooks had only Rustic Bunting but Lewis had Pine Gosbeak, Rustic Bunting and Waxwing. Siberian Jay and Hawk Owl have also been seen here.

8. Kiilopaa is very good for mountain birds. We had 6 Snow Bunting, 3 Ptarmigan and 6 Dotterel here as well as Bluethroat and Dipper lower down. To find it, turn east 1 km south of the road to Kuttura and drive 5 km to the tourist lodge which offers a hotel, youth hostel or camping huts. From here it is an easy 1-2 km walk to the summit. The slopes between here and the next summit proved best for Dotterel.

9. Brooks was successful at a site for White-tailed Eagle found by turning east along a road signposted Sompiojarvi (not Sompiontie), 200 metres North of the bridge at Vuotso. Drive to the lake and scan from there.

10. 35 km North of Sodankyla a boardwalk leads from a car park by the main road to a tower overlooking a marsh. This could be good, especially in the morning (Smith).

Parikkala

Attraction

Was once the only known site in Europe for Blyth's Reed Warbler though this bird can now be found at many sites in both Finland and Sweden. Siikalahti marsh has now been made into a WWF reserve with breeding grebes, swans, ducks, crakes and raptors. This is probably the best known site in Finland for Little Crake; Yellow-breasted Buntings sometimes breed. Nearby are sites for Nutcracker and, in recent years, Booted Warbler.

Notes

1. Siikalahti marsh is an excellent wetland area with Blyth's Reed Warbler, Little Crake, Osprey and many other species. (See page 22.)

2. A nature trail at Likolampi in Parikkala turned out to be unproductive apart from a Golden Oriole. Wherever you can find a vantage point overlooking lakes from Parikkala you could see Red-necked Grebe.

3. Punkaharju is an experimental forestry area with some excellent woodland. A site for Nutcracker and, in the past, Two-barred Crossbill. (See page 24.)

4. Koitsanlahti has been the summer home of a male Booted Warbler every year from 1989-91. (See page 25.)

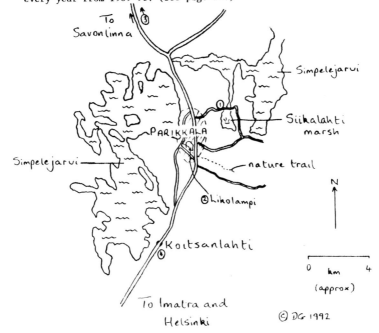

Siikalahti

Getting there

Arriving from the north, ignore the right hand turning into Parikkala village and follow the main road for a further 1.5 km. The marsh is signposted at the next left hand turn towards Kaukola. 2 km from this junction a track to a car park appears suddenly on the right opposite a sign for the reserve.

Notes

1. Around the car park we heard both Thrush Nightingale and Scarlet Rosefinch.

2. 100 metres from the car park is an information centre and picnic area where freelance camping is possible if you can stand the mozzies. Here we had Golden Oriole and Thrush Nightingale and could hear Spotted Crake and Bittern calling from the marsh.

3. An evening walk to the tower along the board walk should certainly produce Spotted Crakes at very close range though virtually impossible to see.

4. The bird tower is excellent especially in the evening when several Hobbies chase around as well as Osprey and Marsh Harrier. Slavonian Grebes are easy to see here, Red-necked Grebes harder. A list in the Information Centre suggested the following breeding numbers (in pairs): Great Crested Grebe (10), Red-necked Grebe (3), Slavonian Grebe (10), Bittern (5), ducks (120+ including Pintail and Garganey), Marsh Harrier (2-4), Water Rail (20), Spotted Crake (50), Little Crake (1-2), Black-headed Gull (1000), Sedge Warbler (600), Reed Warbler (10), Great Reed Warbler (3-6), Rosefinch (10+), Thrush Nightingale (50), Blyth's Reed Warbler (5), Marsh Warbler (1-3), Yellow-breasted Bunting (1-3). We couldn't find the last 2 species (and neither could anyone else judging by the logbook!); the Wardens suggested they weren't present in 1991.

5. This causeway is a good spot to watch for Ospreys. Better still, the next set of bushes on the right (coming from Parikkala) had a singing Blyth's Reed Warbler every night.

6. The site for Little Crake can be found as follows. After the causeway follow the road around to the right, down the east side of the lake. The third track to the right (opposite a small wood and before a green-roofed building) leads down to a red barn which in 1991 had the wreck of a blue lorry outside it. Park here and walk down the track to the pumping station (just a small concrete(?) block). From here, cross the channel and walk left along an embankment until you reach the end of a copse of birches. Listen here around midnight in the marsh opposite three buildings. We also had Blyth's Reed Warbler near the red barn, four Hobbies near the pump station and several calling Spotted Crakes.

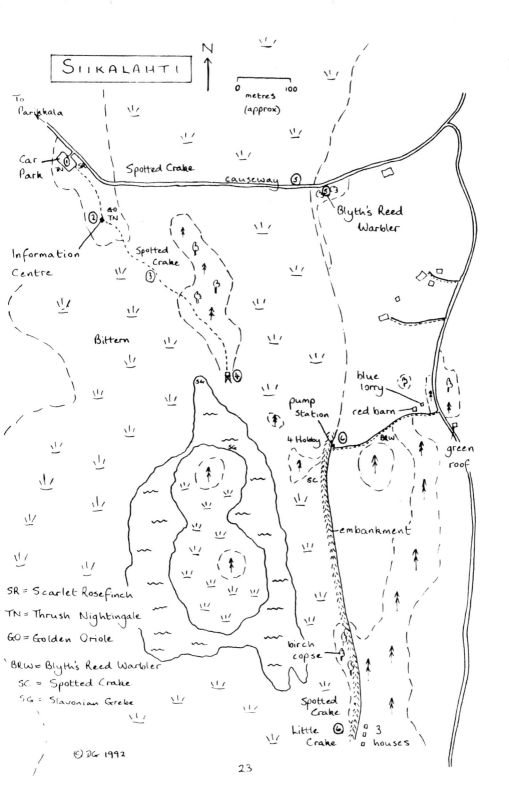

Punkaharju

Getting there

From Parikalla follow the road towards Savonlinna for about 20 km until you see an impressive old railway station "Asema-ateljee" on your right. Just past here, a track signposted "Puistometsaalue", leads off to the right, opposite the new research station. Park on your left by the old research centre.

Notes

1. From the North of the car park, a track running parallel with the road marks the beginning of a nature trail through good-looking woodland. Hazelhen was the only good bird I had here although Black Grouse, Capercaillie and Woodpeckers are also said to occur.

2. Continuing along the main track through the reserve you come to an avenue of lollipop-shaped pines just past a red building on the left. This is one place where the Nutcrackers have been seen. Lewis located them "in Arolla pines". I don't know what these are (and neither did the staff of the forestry centre!) but I assume they are the same as these Swiss Stone Pines (Pinus cembra).

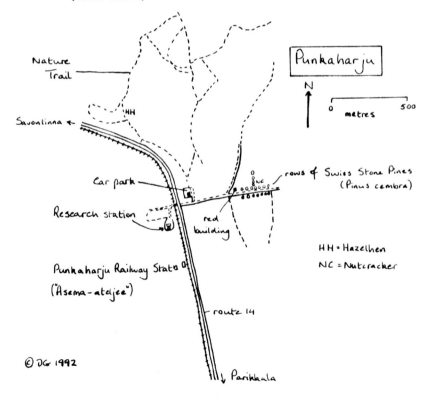

© DG 1992

Koitsanlahti

Getting there

Continue south on the main E6 road from Parikkala. After about 8 km (5 miles) look out for a "K" supermarket on your right before the turn off to "Camping Hovinpuistu". Park in the bus stop just past the shop and look/listen in the field opposite.

Notes

A Booted Warbler has returned to this same field in 1989, 1990 and 1991, singing its heart out every night in search of a mate. We were told that the best time to locate him was between 3.00 and 7.00 a.m. Just past the bus stop is a weakly defined "tractor track" into the field. We walked down this track and located the bird about 20 metres to our left, 60-80 metres from the road. Although difficult to locate in the tall vegetation he gave excellent views. The song was delivered in short but frequent bursts like an extended Whitethroat song but softer, livelier and with a tinkling quality.

A Scarlet Rosefinch was also here, singing a more elaborate song than usual.

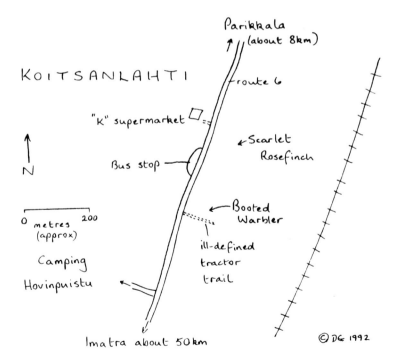

Jyvaskula

Attraction

In 1991, Great Grey Owl, Ural Owl, Tengmalm's Owl and Pygmy Owl were all found nesting here although it is unusual to have Great Greys so far south. River Warbler and Greenish Warbler can also be found here in spring.

Getting there

Jyvaskula is in the middle of southern Finland, 351 km south of Oulu and 277 km north of Helsinki.

Notes

I cannot divulge the nest sites of most of the owl species although the local birder who found it has given me permission to describe the Pygmy Owl site.

1. Follow the road towards Saukkola from Jyvaskula and turn sharp right by a spectacular old wooden building just before reaching Saukkola (signposted Riskopera, I think). Follow this track, basically keeping left (but don't take the track to the farmhouse) until a lake appears on your left. Park here. The Pygmy Owl nested between the road and the lake. Further back, opposite the farmhouse, a pair of Long-eared Owls hunted around the open fields.

2. A site for River Warbler is as follows. Drive north-east along route 637 towards Laukaa. Don't turn off into Laukaa itself but follow the main road as it bends west. About a kilometre after passing Laukaa look out for a bridge over a railway. Park here and listen for River Warbler to the north of the road and east of the railway.

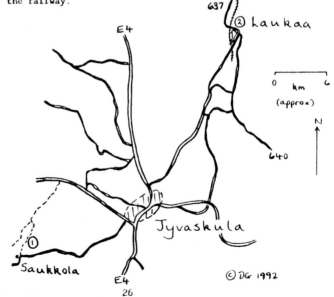

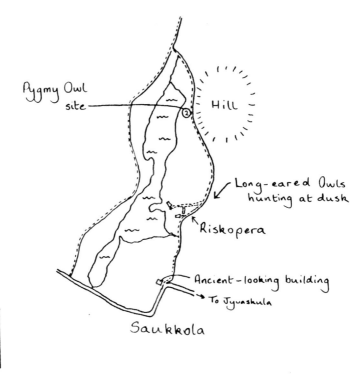

Sites near Jyvaskula

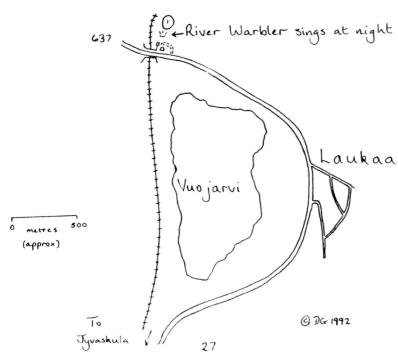

Sammalinen and Vakevala

Attraction

This area is a migration watchpoint for locating aerial migrants flying to/from the Arctic (Barnacle Geese etc.). However, such studies have also revealed the regular presence of a pair of Spotted Eagles which wander over to here from their breeding site in Russia.

Getting there

Drive south on the 387 from Lappeenranta towards Virolahti and Vaalimaa. The two vantage points are to the east of this road after about 40 km.

Notes

The two vantage points can be found as follows:

1. About 6 km south of Ylamaa look out for a sign to Sammalinen. Follow this road for about 2 km and look for a lay-by on your right below a ridge half-hidden by trees. Park there and find a small path through the trees, over the rocks to the top of the ridge. From here is an impressive view in almost every direction. Behind you you should find a green plastic box on a post. The log book in there makes fascinating reading and includes superb sketches of the eagle by Dan Zetterstrom (or is it Hakan Delin?). Only one person, an American, had reported a Spotted Eagle here in 1991. We had a few raptors including Buzzard, Honey Buzzard, Goshawk and Osprey but no eagles.

2. Return to the main 387 road and drive south for a further 2.0 miles to Hujakkala. Turn left here to Vakevala and follow that road for 4.7 miles keeping right, with a lake on your right until you reach an obvious wooden bridge with a marsh to your left. Park and scan from here, looking east towards the Soviet Union. Frankly there wasn't much of a viewpoint from here although Spotted Eagles do like lakes and marshes so if one did appear you'd get great views. We only had a selection of ducks and waders on the marsh to entertain us.

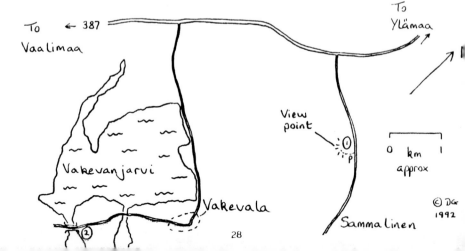

Finland up-date 1992

Most of these sightings were made between 5th and 30th June 1992 by D. Gosney and a Gostours party with additional information from Greg Baker.

Oulu Oilport (page 4)

The roadside pools had 2 pairs of Temminck's Stints but no Terek Sandpipers although we did see these by gaining access to a private area nearby.

Kerseleenlahti (page 5)

Again we couldn't find Yellow-breasted Bunting here although a noticeboard confirmed that this is a known site for them.

Liminka tower (page 6, site 1)

This tower was excellent for birds, especially Little Gulls and Cranes, and is reasonably accessible as long as you don't follow signs to Lintuntuuri, as previously suggested. A much better route, by car, is as follows: Drive North through Liminka, continue past the sign to Lintuntuuri and instead take the next left. Follow this road for about 3 km to a T-junction. Turn left here and then, as the road bends to the right, look for a track which continues straight on. Follow this track for almost 2 km to a collection of holiday chalets. From the northern end of these chalets you will see the tower which can be reached via a little footbridge and a poorly-marked, and sometimes muddy, path.

Kuusamo (page 8)

There were hardly any reports of Arctic or Greenish Warblers around Valtavaara in 1992 (and no Two-barred Crossbills). However, we did find Arctic Warbler at another, apparently more reliable, site to the West of Kuusamo (see map overleaf). Another good area was at Ahvenjarvi, an excellent spot for feeding waders such as Ruff, Wood Sandpiper and Spotted Redshank. We also had a Great Grey Shrike here.

Valtavaara (page 11)

Up to 3 singing Red-flanked Bluetails occurred in 1992, of which we saw one, a splendid blue adult, on Kontainen. Its song, given from 03.30 to 05.30 was nothing like my previous description (based on tapes). Every 15-60 seconds it repeated a simple ditty 'syoo-sissup, swee-sissup' more predictable than both Redwing and Redstart and lacking their 'dribbly' endings. Another bonus was a Tengmalm's Owl in daylight near the 'crossbill puddle' (which was dry).

Pyhatuntuuri (page 15)

'Extra' species here included a Black Woodpecker posing near the ski-lift, a pair of Rustic Buntings just West of the observation tower and a singing Greenish Warbler near the southernmost set of wooden steps. On 12/13th July, Greg Baker added Hazelhen (2 places), Black Grouse and Wryneck and had a Greenish Warbler higher up the slope.

Karigasniemi (page 16)

Mount Ailigas produced 3 Dotterel and 3 Ptarmigan on the stony areas around the tower, plus a total of 8 Snow Bunting and a pair of Long-tailed Skuas (on the slopes between the tower and the barrier). At the marsh at 13 km, we again had Broad-billed Sandpiper and Jack Snipe as well as excellent views of Arctic Redpoll.

Ivalo (page 18)

Site 3b is indeed the site for Jack Snipe (Lewis) but a better site for Little Bunting is found as follows. In Ivalo, follow signs to Murmansk until, after a few km, you see an 80-km speed-limit sign. Park here and walk for 100-200 metres, looking on both sides of the road. Little Buntings were obvious here even though they'd been most elusive at Kuusamo.

Kilopaa (page 20)

A bonus here was a Gyr Falcon which flew up 50 metres in front of me on the southern slope.

Siikalahti (page 22)

Blyth's Reed Warblers were very elusive. We were told of another good spot for them: the car-park at the end of the northerly track which begins opposite the reserve car-park.

Punkaharju (page 24)

Nutcrackers performed beautifully, especially by the track to the South of the red building, where we also had Hazelhen and Ortolan Bunting.

Koitsanlahti (page 25)

The Booted Warbler returned again, but our dawn visit produced 'only' Blyth's Reed, Marsh and Icterine Warblers.

Jyvaskula (page 26)

Ural, Pygmy, Tengmalm's and Long-eared Owls all gave excellent views and we had Hazelhens at two sites but the River Warbler wasn't singing when we visited.

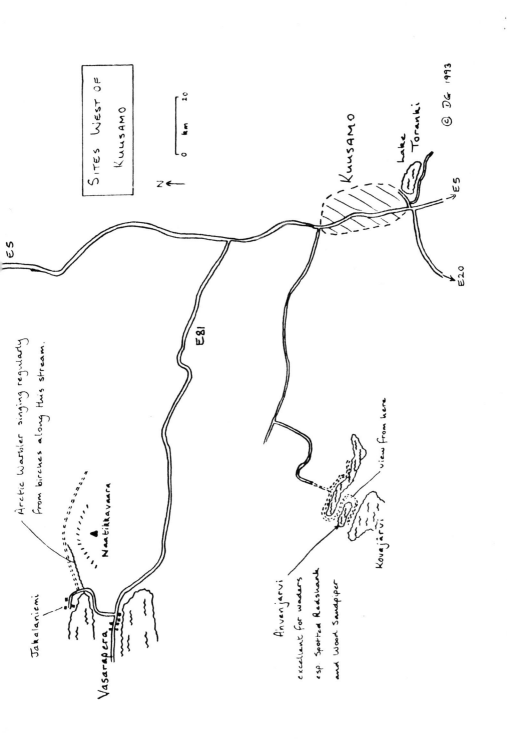

Sammalinen (page 28)

The Spotted Eagle was definitely seen in 1992 but, like most visitors, our best bird here was a Black Stork.

Sodankyla (South of page 18)

A local birdwatcher, Ossi Pihajoki, is keen to practise his english by helping British birdwatchers to see owls and woodpeckers. In late June he did well to find us a Three-toed Woodpecker but if you can arrive before mid-June, the owls should still be around their nest-sites. Ossi can be contacted by ringing 9693 610099

Finland update 1993

Stuart and Sue Burnet found the following additional species in mid-June: a male Capercaillie at Juntaan Luonnonkoitomesta, Arctic Redpoll at Kilopaa and Great Grey Shrike on the Kuttura road (all page 18), Pine Grosbeak and Hazelhen near Lake Toranki (page 10) and 2 Caspian Terns near Oulu Oilport (page 4). They also had 'huge numbers' of Caspian Terns by driving along a causeway to the town of Seskaro, at the head of the Gulf of Bothnia.
Another bonus was a dispaying Olive-backed Pipit in a birch wood near Oulu where they also found Rustic and Little Buntings. Other birders they met had seen both Red-flanked Bluetail and Two-barred Crossbill at Valtavaara.

Finland up-date 1994

A flying visit by myself and Max Whitby produced most of the expected species. Despite it being 'the worst year for rodents since 1975' we still had Hawk Owl, Great Grey Owl and Eagle Owl near Sodankyla where other species included Little Bunting, Siberian Tit, Siberian Jay, Goshawk, Three-toed Woodpecker, Velvet Scoter, Red-necked Phalarope, Smew and Whimbrel. At Kuusamo we had more joy with Little Bunting and Rustic Bunting by driving to the far end of the Lake Toranki road and listening from there. More superb views of Arctic Redpoll at Karigasniemi marsh (13 km) suggests that this is a reliable site for them, although there are Mealy's there too. An easier way to find Oulu Oilport is to follow signs to Nuottasari from the main E4 road - but we didn't see Terek Sandpipers anyway.
Almost all the best birds we had, both here and on Varanger Fjord, are superbly captured in our video entitled 'Gosney in the Arctic' - see the inside back cover for more details.

Your chances of finding birds around Kuusamo are greatly increased if you go after the Kuusamo Bird Race (it was on 17th June in1994). This competition is taken very seriously and Finnish birders are reluctant to divulge any 'sites' until the race is over. In 1994, there were no Bluetails on Valtavaara although at least one had been found in Oulanka National Park.